MILITARY SUBMARINES

Martha London

DiscoverRoo
An Imprint of Pop!
popbooksonline.com

abdobooks.com

Published by Pop!, a division of ABDO, PO Box 398166, Minneapolis, Minnesota 55439. Copyright © 2020 by POP, LLC. International copyrights reserved in all countries. No part of this book may be reproduced in any form without written permission from the publisher. Pop!™ is a trademark and logo of POP, LLC.

Printed in China.

052019
092019

THIS BOOK CONTAINS RECYCLED MATERIALS

Cover Photo: Shutterstock Images
Interior Photos: Shutterstock Images, 1, 6, 8–9, 14, 31; US National Archives and Records Administration, 5, 13, 20, 21, 30; Arkivi/Hulton Archive/Getty Images, 7; Library of Congress, 11, 24 (top); Niday Picture Library/Alamy, 12; US Navy, 15, 19, 27, 28, 29; Scherl/Süddeutsche Zeitung Photo/Alamy, 16; iStockphoto, 17; Dimitri Newman/Alamy, 22; Naval History and Heritage Command, 23, 25 (top); charistoone-images/Alamy, 24 (bottom); Shen Cheng/Xinhua News Agency/Newscom, 25 (bottom)

Editor: Connor Stratton
Series Designer: Jake Slavik
Library of Congress Control Number: 2018964858
Publisher's Cataloging-in-Publication Data

Names: London, Martha, author.
Title: Military submarines / by Martha London.
Description: Minneapolis, Minnesota : Pop!, 2020 | Series: Inside the military | Includes online resources and index.
Identifiers: ISBN 9781532163876 (lib. bdg.) | ISBN 9781644940600 (pbk.) | ISBN 9781532165313 (ebook)
Subjects: LCSH: Submarine boats--Juvenile literature. | Submarine warfare--Juvenile literature. | Naval operations--Submarine--Juvenile literature. | Underwater navigation--Juvenile literature.
Classification: DDC 359.83--dc23

WELCOME TO DiscoverRoo!

Pop open this book and you'll find QR codes loaded with information, so you can learn even more!

Scan this code* and others like it while you read, or visit the website below to make this book pop!

popbooksonline.com/military-submarines

*Scanning QR codes requires a web-enabled smart device with a QR code reader app and a camera.

TABLE OF CONTENTS

CHAPTER 1

MILITARY SUBMARINES IN ACTION

In February 1945, a German submarine

dived near the coast of Norway.

A British submarine followed behind.

The British captain was trying

to sink the German submarine.

German submarines are known as U-boats.

He followed it for hours. Then he fired

four **torpedoes** at it.

The German
submarine dived
deeper to avoid the
torpedoes. Three
of the torpedoes
missed. But the fourth
torpedo went deeper
than the others.
It made a direct
hit. The German
submarine sank.

The only submarine to survive an underwater submarine battle was named HMS Venturer.

A submarine battle happened

underwater only once in history.

But many navies have used submarines.

These ships can move below water. They

are hard to spot. As a result, militaries

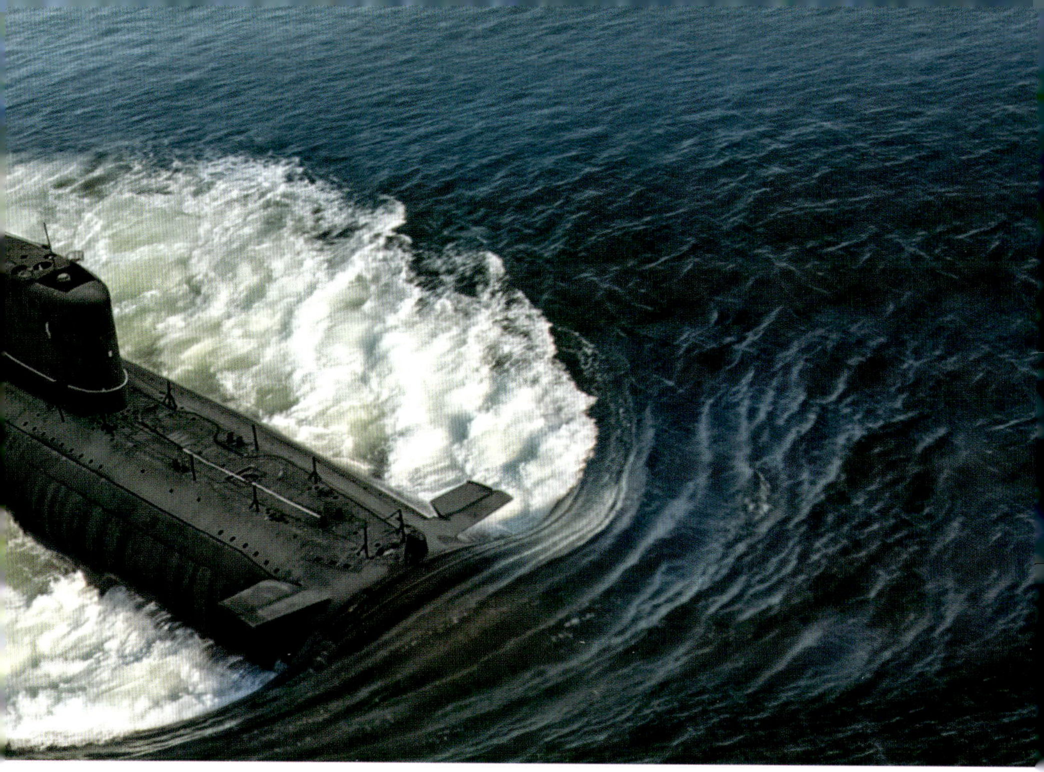

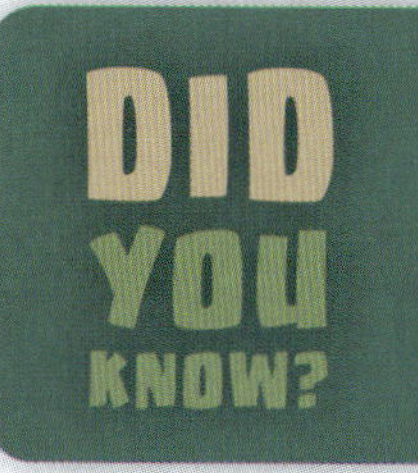

In 2018, the US Navy had 71 active submarines. This fleet was one of the largest in the world.

can use submarines to spy. Submarines can also sneak up on enemy ships.

DID YOU KNOW?

Submarines fought one another as early as World War I (1914–1918).

CHAPTER 2

HISTORY OF MILITARY SUBMARINES

People worked for many years to create underwater vehicles. In 1620, a Dutch inventor built the first working submarine. Until 1800, some subs used oars to move. Others used **propellers**.

A drawing shows a submarine built in the early 1800s.

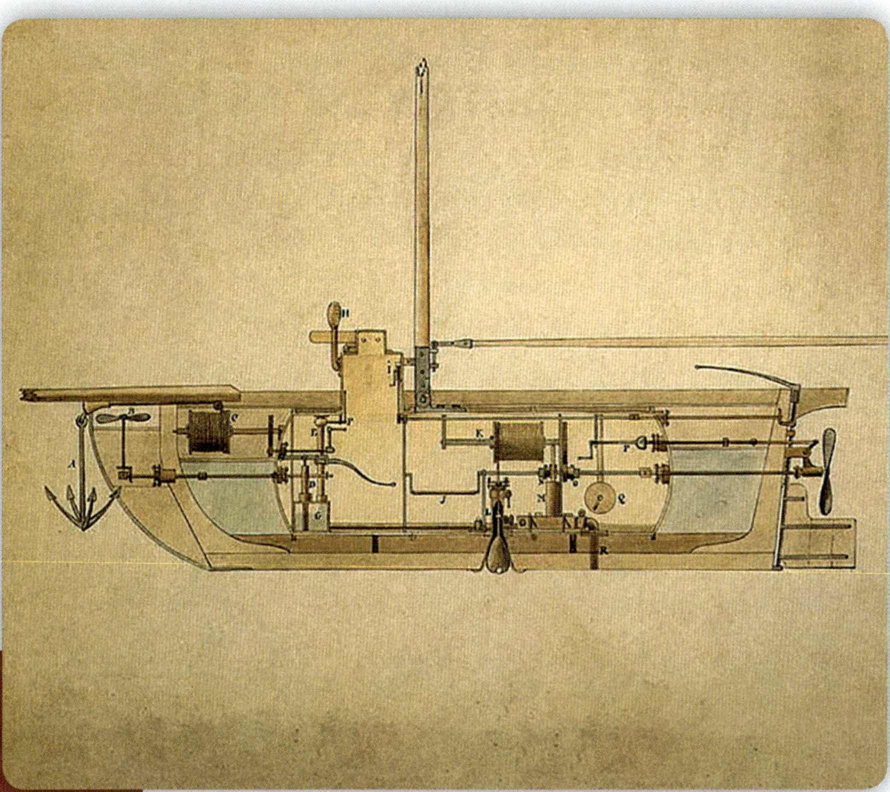

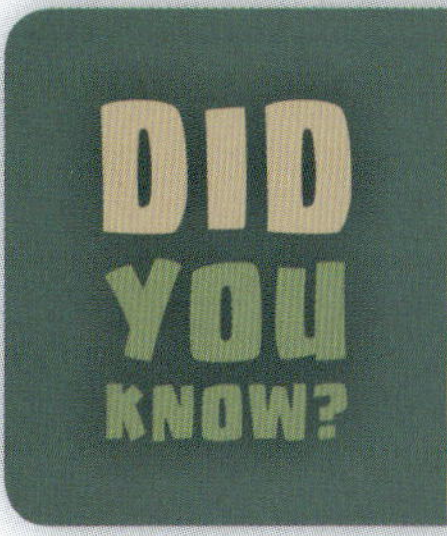

DID YOU KNOW?

In the early 1800s, an inventor built a submarine with a propeller. The sub also had a sail to power it above water.

The first electric-powered submarine was the USS Holland.

In 1900, the US Navy used the first electric-powered submarines. The subs used **batteries** and engines for power.

Around this time, navies also started loading submarines with **torpedoes**.

In the 1910s, militaries began building submarines with **sonar**. Sonar devices send out pulses of sound. The sound hits an object. Then the sound bounces back. The echoes show that an object is nearby.

A US sailor studies a sonar screen. These screens show if the sonar device has found anything.

A German submarine snorkel pokes out of the ocean during World War II.

Early submarines could only stay underwater as long as their batteries were charged. Subs rose to the surface to recharge. Even during World War II (1939–1945), submarines could remain underwater for only a couple days.

PARTS OF A SUBMARINE

After World War II, new submarines could stay underwater for a long time. In 1954, the US Navy built a sub that used **nuclear** energy. This ship could spend months underwater at one time.

DID YOU KNOW?

Early submarines moved slowly underwater before attacking enemies. Otherwise, their **batteries** would die. Nuclear subs can move at top speed underwater for months.

A US sailor wears a mask that gives him air to breathe.

Nuclear submarine crews needed

air they could breathe underwater.

They also needed fresh water to drink.

Navies solved these problems. Air tanks helped crews breathe. And machines on board took the salt out of ocean water.

A US sailor checks a machine on a nuclear submarine. This machine makes the air easier to breathe by taking away carbon dioxide.

In the 1960s, navies loaded
submarines with nuclear weapons. One
nuclear weapon can destroy a large
city. Navies did not want enemies to
know where these weapons were. They
made submarines quieter. In the
1980s, navies built subs with
smoother **propellers**.
These propellers made
less sound in the water.
Quieter submarines
are harder for
sonar to notice.

The first US submarine to carry nuclear weapons was the USS George Washington.

TIMELINE

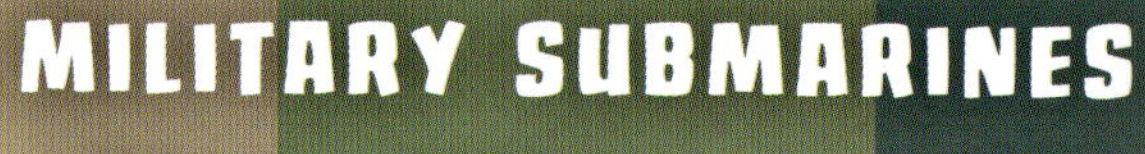

1866

A British inventor builds the modern **torpedo**.

1620

A Dutch inventor makes the first working submarine.

1898

A US inventor builds the first electric-powered submarine.

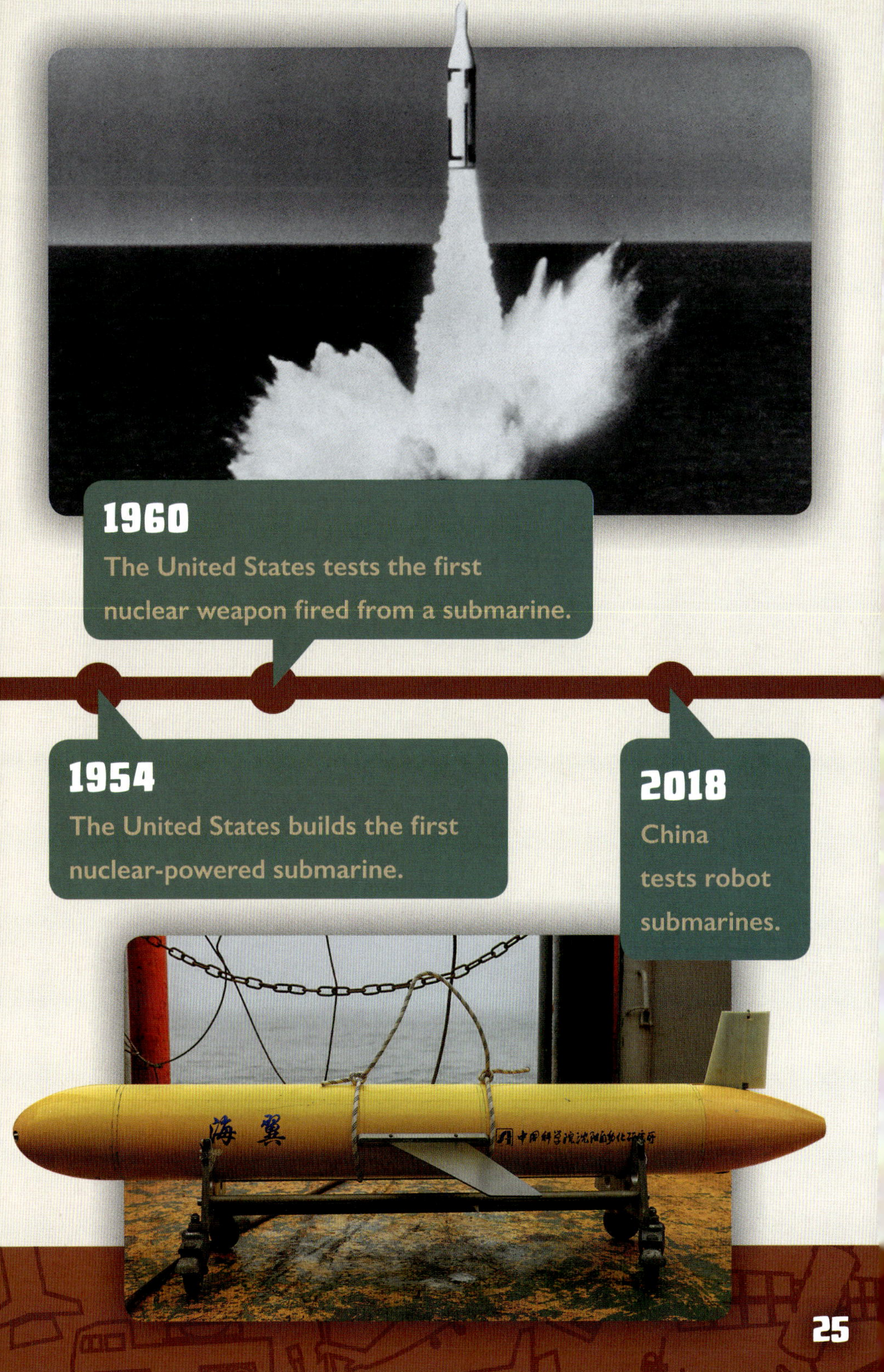

1960
The United States tests the first nuclear weapon fired from a submarine.

1954
The United States builds the first nuclear-powered submarine.

2018
China tests robot submarines.

CHAPTER 4

MILITARY SUBMARINES TODAY

Today's submarines are hard for **sonar** to notice. Some navies cover subs with a layer of air bubbles. Water hitting air is quieter than water hitting metal. These submarines may hide better from sonar.

Submarines covered with air bubbles are faster than submarines without these layers.
air bubble layer

Scientists are also working on robot submarines. These ships have no humans on board. Instead, soldiers control them from land. A robot sub can be smaller than other subs. The ship's size makes it harder for enemies to find.

A soldier steers a robot submarine with a remote control.

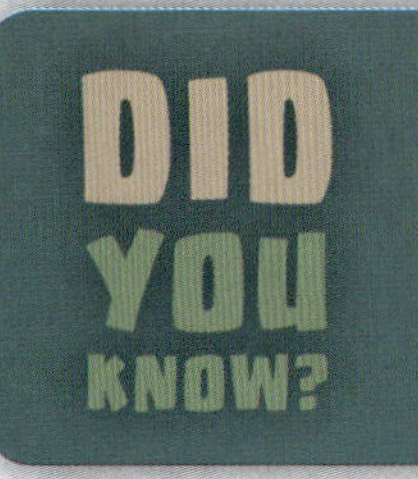

Some robot submarines are built to look like fish.

DID YOU KNOW?

In 2017, the US Navy started its first unit for robot submarines.

MAKING CONNECTIONS

TEXT-TO-SELF

Would you like to travel in a submarine? Why or why not?

TEXT-TO-TEXT

Have you read books about other military vehicles? What do those vehicles have in common with submarines?

TEXT-TO-WORLD

Military submarines are often used to explore areas. Why would traveling underwater be helpful for doing this task?

GLOSSARY

battery – a device that stores electricity and can be used to power certain items.

nuclear – relating to a type of energy made by breaking atoms apart.

propeller – a set of blades that spin and help things move.

snorkel – a submarine tube that pokes out above water, letting the engine work while the submarine is underwater.

sonar – a system that locates things by bouncing sound waves off them.

torpedo – an explosive weapon that is shot through the water.

INDEX

ONLINE RESOURCES

popbooksonline.com

Scan this code* and others like it while you read, or visit the website below to make this book pop!

popbooksonline.com/military-submarines

*Scanning QR codes requires a web-enabled smart device with a QR code reader app and a camera.